ISBN 978-0-365-68450-3
PIBN 11255544

1 MONTH OF
FREE
READING

at

www.ForgottenBooks.com

By purchasing this book you are eligible for one month membership to ForgottenBooks.com, giving you unlimited access to our entire collection of over 1,000,000 titles via our web site and mobile apps.

To claim your free month visit:
www.forgottenbooks.com/free1255544

NBS SPEC. PUBL. **260-19**

Standard Reference Materials:

ANALYSIS OF INTERLABORATORY MEASUREMENTS ON THE VAPOR PRESSURE OF GOLD
(CERTIFICATION OF STANDARD REFERENCE MATERIAL 745)

U.S. Department of Commerce

National Bureau of Standards

NATIONAL BUREAU OF STANDARDS

The National Bureau of Standards [1] was established by an act of Congress March 3, 1901. Today, in addition to serving as the Nation's central measurement laboratory, the Bureau is a principal focal point in the Federal Government for assuring maximum application of the physical and engineering sciences to the advancement of technology in industry and commerce. To this end the Bureau conducts research and provides central national services in four broad program areas. These are: (1) basic measurements and standards, (2) materials measurements and standards, (3) technological measurements and standards, and (4) transfer of technology.

The Bureau comprises the Institute for Basic Standards, the Institute for Materials Research, the Institute for Applied Technology, the Center for Radiation Research, the Center for Computer Sciences and Technology, and the Office for Information Programs.

THE INSTITUTE FOR BASIC STANDARDS provides the central basis within the United States of a complete and consistent system of physical measurement; coordinates that system with measurement systems of other nations; and furnishes essential services leading to accurate and uniform physical measurements throughout the Nation's scientific community, industry, and commerce. The Institute consists of an Office of Measurement Services and the following technical divisions:

> Applied Mathematics—Electricity—Metrology—Mechanics—Heat—Atomic and Molecular Physics—Radio Physics [2]—Radio Engineering [2]—Time and Frequency [2]—Astrophysics [2]—Cryogenics.[2]

THE INSTITUTE FOR MATERIALS RESEARCH conducts materials research leading to improved methods of measurement standards, and data on the properties of well-characterized materials needed by industry, commerce, educational institutions, and Government; develops, produces, and distributes standard reference materials; relates the physical and chemical properties of materials to their behavior and their interaction with their environments; and provides advisory and research services to other Government agencies. The Institute consists of an Office of Standard Reference Materials and the following divisions:

> Analytical Chemistry—Polymers—Metallurgy—Inorganic Materials—Physical Chemistry.

THE INSTITUTE FOR APPLIED TECHNOLOGY provides technical services to promote the use of available technology and to facilitate technological innovation in industry and Government; cooperates with public and private organizations in the development of technological standards, and test methodologies; and provides advisory and research services for Federal, state, and local government agencies. The Institute consists of the following technical divisions and offices:

> Engineering Standards—Weights and Measures — Invention and Innovation — Vehicle Systems Research—Product Evaluation—Building Research—Instrument Shops—Measurement Engineering—Electronic Technology—Technical Analysis.

THE CENTER FOR RADIATION RESEARCH engages in research, measurement, and application of radiation to the solution of Bureau mission problems and the problems of other agencies and institutions. The Center consists of the following divisions:

> Reactor Radiation—Linac Radiation—Nuclear Radiation—Applied Radiation.

THE CENTER FOR COMPUTER SCIENCES AND TECHNOLOGY conducts research and provides technical services designed to aid Government agencies in the selection, acquisition, and effective use of automatic data processing equipment; and serves as the principal focus for the development of Federal standards for automatic data processing equipment, techniques, and computer languages. The Center consists of the following offices and divisions:

> Information Processing Standards—Computer Information — Computer Services — Systems Development—Information Processing Technology.

THE OFFICE FOR INFORMATION PROGRAMS promotes optimum dissemination and accessibility of scientific information generated within NBS and other agencies of the Federal government; promotes the development of the National Standard Reference Data System and a system of information analysis centers dealing with the broader aspects of the National Measurement System, and provides appropriate services to ensure that the NBS staff has optimum accessibility to the scientific information of the world. The Office consists of the following organizational units:

> Office of Standard Reference Data—Clearinghouse for Federal Scientific and Technical Information [3]—Office of Technical Information and Publications—Library—Office of Public Information—Office of International Relations.

---

[1] Headquarters and Laboratories at Gaithersburg, Maryland, unless otherwise noted; mailing address Washington, D.C. 20234.
[2] Located at Boulder, Colorado 80302.
[3] Located at 5285 Port Royal Road, Springfield, Virginia 22151.

Standard Reference Materials:

Analysis of Interlaboratory Measurements on the Vapor Pressure of Gold (Certification of Standard Reference Material 745)

Robert C. Paule and John Mandel

Institute for Materials Research
National Bureau of Standards
Washington, D.C. 20234

t.

U. S, National Bureau of Standards, Special Publication 260–19

Nat. Bur. Stand. (U.S.), Spec. Publ. 260–19, 21 pages (January 1970)

CODEN: XNBSA

Issued January 1970

No.250

Library of Congress Catalog Card Number: 70-603290

Contents

Tables

Figures

Analysis of Interlaboratory Measurements on the Vapor Pressure of Gold (Certification of Standard Reference Material 745)

Robert C. Paule and John Mandel

A detailed statistical analysis has been made of results obtained from a series of interlaboratory measurements on the vapor pressure of gold. The gold Standard Reference Material 745 which was used for the measurements has been certified over the pressure range 10^{-8} to 10^{-3} atm. The temperature range corresponding to these pressures is 1300–2100 K. The gold heat of sublimation at 298 K and the associated standard error were found to be $87,720 \pm 210$ cal/mol ($367,040 \pm 900$ J/mol). Estimates of uncertainty have been calculated for the certified temperature-pressure values as well as for the uncertainties expected from a typical single laboratory's measurements. The statistical analysis has also been made for both the second and third law methods, and for the within- and between-laboratory components of error. Several notable differences in second and third law errors are observed.

Key words: Components of error (within- and between-laboratories); gold; heats of sublimation (second and third law); interlaboratory measurements; standard errors; standard reference materials; vapor pressure.

1. Introduction

This report is part of a program to establish five standard reference materials. The materials, Cd, Ag, Au, Pt, and W, are being certified by the National Bureau of Standards for their vapor pressures as a function of temperature. Certification covers the 10^{-8} to 10^{-3} atm range. For the complete series of materials, the temperatures corresponding to the above pressures will range from 600 to 3000 K. Gold, the first material to be certified, covers a temperature range from 1300 to 2100 K. The gold standard reference material is now available for sale to the public [1].[1]

Experience in high-temperature vapor-pressure measurements has shown that large systematic errors in pressure of 30, 50, or even 100 percent are not uncommon, even among experienced investigators. The vapor pressure standard reference materials will allow workers in the field to detect such systematic errors and to evaluate the precision and accuracy of their measurements. The materials should be most useful for checking low vapor pressure measurement methods such as the Knudsen, torque Knudsen, Langmuir, and mass spectrometric methods.

This report will give estimates of the uncertainty of the certified temperature-pressure values as well as estimates of the uncertainties of a "typical" single laboratory's measurements. These uncertainties summarize results obtained from interlaboratory tests made in 1968 (see list of cooperative laboratories). The uncertainties represent current practice and should not be considered fixed with respect to time and progress. We believe the uncertainties will be reduced in the future through the use of the vapor pressure standard reference materials.

---

[1] Figures in brackets indicate footnotes and references beginning on p. 7.

The results from the 1968 interlaboratory tests were used to obtain a composite heat of sublimation for gold at 298 K. The certified temperature-pressure values were then obtained by back-calculating through the third law equation

$$T\left[\Delta\left(-\frac{G_T^\circ - H_{298}^\circ}{T}\right) - R \ln P\right] = \Delta H_{\text{sub}298} \quad (1)$$

using the composite $\Delta H_{\text{sub}298}$ of 87,720 cal/mol [2] (367,040 J/mol) [3] and the referenced free energy functions [4]. P is expressed in atmospheres. All temperatures for this report have been converted to the 1968 International Practical Temperature Scale (IPTS–68). The certified temperature-pressure values as well as the corresponding $1/T$ and $\log P$ values are listed below.

T(K)	P(atm) [3]	$(1/T) \times 10^4$ (K^{-1})	$\mathrm{Log}_{10}P$(atm)[3]
1300	9.92×10^{-9}	7.692	-8.003
1338 (M.P.)	2.56×10^{-8}	7.474	-7.592
1400	1.01×10^{-7}	7.143	-6.993
1500	7.36×10^{-7}	6.667	-6.133
1600	4.14×10^{-6}	6.250	-5.383
1700	1.90×10^{-5}	5.882	-4.721
1800	7.25×10^{-5}	5.556	-4.139
1900	2.42×10^{-4}	5.263	-3.616
2000	7.07×10^{-4}	5.000	-3.151
2100	1.87×10^{-3}	4.762	-2.727

A broad cross-section of measurement techniques were used by the cooperating laboratories in the interlaboratory tests; the techniques included the Knudsen (weight loss and condensation methods), torque Knudsen, and calibrated mass spectrometric methods. Summary information regarding the experimental details for each laboratory are given in table 1.

TABLE 1. *Summary of Experimental Methods*

Labo-ratory	Method	Temperature measurement technique	Container material	Effective orifice area × 10³, cm²	Remarks
1	Knudsen using conden-sation plates	Optical pyrometer with black-body hole	W crucible with graphite or Al₂O₃ insert cups	2.84, 11.59	In preliminary experi-ments Au wet and crept excessively on bare W cell
2	Knudsen using conden-sation plates and x-ray fluorescence detection	Optical pyrometer with black-body hole	Mo cell	0.70, 2.15, 5.85	Au wet Mo and some creep noted
3	Knudsen	Pt-Rh, Pt thermocouple	Quartz; pyrolytic graphite	0.36, 2.60, 3.91	
4	Knudsen	Optical pyrometer with black-body hole	Pyrolytic graphite	3.21, 4.93	
5	Knudsen plus Knudsen cell in mass spectrom-eter	Optical pyrometer	Carbon	1.53	First Knudsen results used to calibrate mass spectrometer
6	Knudsen	Pt-13%Rh, Pt thermo-couple	ZTA graphite	62.4	
7	Torque Knudsen	Pt-13%Rh, Pt thermo-couple	ZTA graphite	8.5, 31.9, 55.7	Each curve's temperature measurements made in one direction only
8	Torque Knudsen	Pt-10%Rh, Pt and Pt-30%Rh, Pt-6%Rh thermocouples	Pyrolytic graphite	1.36, 9.40	Each curve's temperature measurements made in one direction only
9	Knudsen cell in mass spectrometer with absolute weight cali-bration	Optical pyrometer look-ing at side of crucible	Ir crucible with Al₂O₃ insert cup	4.7	Au₂ also measured
10	Triple Knudsen cell in TOF mass spectrom-eter using Ag and Hultgren Ag vapor pressure values and Mann's cross sections for calibration	Optical pyrometer with black-body hole	W crucible	0.49	
11	Double oven Knudsen cell used for absolute Au calibration of TOF mass spectrometer	Optical pyrometer sight-ing into orifice	Graphite		

2. Treatment of Data

The detailed temperature-pressure data from the 11 laboratories which measured the vapor pressure of gold are given in table 2 (see sec. 6.3). Plots of the data are given in figures 1 through 5 of section 6.3. The solid line in these figures represents the pooled curve for all accepted data from all laboratories. A total of 41 sets of data (runs) with over 350 temper-ature-pressure points were available for considera-tion. Each temperature-pressure run has been used to obtain both the second and third law heats of sublimation at 298 K. Equation (1) was used to calculate the individual third law ΔH_{sub298} values corresponding to each temperature-pressure point and the average ΔH_{sub298} value was calculated for

each run. The evaporation coefficient for gold has been assumed to be unity. In agreement with this assumption, we observed no evidence of trend in third law heats with changing orifice area.

The second law heat for each vapor pressure-temperature run was obtained by least-squares fitting the A and B constants in the equation:

$$\Delta\left(-\frac{G_T^\circ - H_{298}^\circ}{T}\right) - R \ln P = A + \frac{B}{T} \qquad (2)$$

where P is expressed in atmospheres. This calcula-tional procedure is similar to the sigma method, and does not require the specification of a mean effective temperature [5]. The slope B is the second law heat of sublimation at 298 K. The intercept A

2

will be zero for the ideal case where the measured pressures and the free energy functions are completely accurate. We have kept the intercept A in the least-squares equation to allow for possible error. This second law procedure is very convenient to use when the calculations, including the interpolation of free energy functions, are made by computer. The OMNITAB computer language [6] was used in this work. A summary of the second and third law results is given in tables 3 and 4.

TABLE 3. *Summary of Second Law Results*

Lab No.	Run No.	No. of points	Intercept A (see eq. 2) cal · mol⁻¹ · deg⁻¹	Slope B, 2d Law ΔH_{sub296} (see eq. 2) cal · mol⁻¹	f_1 (see eq. 7)	f_2 (see eq. 8)	S_{fit} (see eq. 5)
1	1	11	−.744	89664	4.73	8510	.1904
	2	12	−.957	90003	3.77	6570	.2209
2	1	10	.698	87176	7.95	14190	.1364
	2	9	−2.003	91206	5.99	10650	.1039
	4.	6	−2.809	92638	15.57	25820	.1224
3	1	11	1.080	85924	9.05	15580	.0771
	2	10	1.906	84868	9.07	15390	.0690
	3	8	−1.640	90142	17.27	28060	.0993
4	1	10	−.531	89133	5.65	10480	.2288
	2	7	1.437	85876	7.20	13420	.1650
	3	9	−1.238	89815	6.11	11310	.1835
5	1	5	−.779	88508	7.31	11880	.0193
	2	31	−.329	87758	2.59	4300	.1194
7	1	10	−.352	88464	11.41	17870	.0559
	2	6	1.833	85015	13.42	21220	.0196
	3	6	1.392	85687	14.44	22740	.0409
	4	12	−1.081	89500	11.61	18340	.0499
	5	6	−.829	88911	11.57	17730	.0463
	6	15	.868	86480	11.52	18100	.0304
	7	7	−1.648	90212	13.41	22130	.0662
8	1	11	−.130	86862	4.95	7900	.0426
	2	11	.128	86300	5.99	10090	.0547
	3	9	−.302	87041	7.35	11450	.0392
	4	10	−.148	86927	5.28	8380	.0562
	5	11	−.236	86931	6.07	10260	.0199
9	1	13	−1.409	87338	3.91	6170	.0706
	2	14	−1.493	87702	3.69	5880	.1681
	3	14	−3.368	91092	3.58	5710	.1275
	4	14	−3.098	90977	3.62	5740	.1971
10	1	6	2.472	81981	13.61	22010	.1190
	2	7	−5.184	98741	11.64	19990	.3271
	3	10	4.746	77563	6.92	11080	.2544
	4	17	.020	84593	5.66	9020	.2350
11	1	6	−6.594	98536	12.96	20850	.0538
	2	5	−7.390	99568	19.61	30560	.3024
	3	6	10.508	70700	12.48	20410	.2270
	4	5	6.400	77335	30.62	49260	.1955
	5	5.	−.947	89209	12.80	20470	.3695

TABLE 4. *Summary of Third Law Results*

Lab No.	Run No.	No. of points	3d Law ΔH_{sub298} cal·mol^{-1}	f_3 (see eq. 10)	S_{fit} (see eq. 9)
1	1	11	88316	.302	325.2
	2	12	88320	.289	393.9
2	1	10	88425	.316	230.2
	2	9	87626	.333	269.4
	3	2	87747	.707	47.6
	4	6	87975	.408	225.7
	5	2	88120	.707	133.6
3	1	11	87786	.302	141.3
	2	10	88108	.316	160.8
	3	8	87477	.354	158.2
4	1	10	88142	.316	412.1
	2	7	88566	.378	318.0
	3	9	87514	.333	345.4
5	2	31	87208	.180	201.5
6	1	4	88068	.500	402.9
7	1	10	87912	.316	82.8
	2	6	87917	.408	99.7
	3	6	87882	.408	87.6
	4	12	87791	.289	85.5
	5	6	87638	.408	79.1
	6	15	87845	.258	55.8
	7	7	87489	.378	128.6
8	1	11	86653	.302	61.0
	2	11	86517	.302	89.3
	3	9	86570	.333	58.4
	4	10	86691	.316	85.0
	5	11	86531	.302	39.4
9	1	13	85097	.277	195.6
	2	14	85304	.267	322.4
	3	14	85679	.267	461.2
	4	14	86016	.267	492.8
10	1	6	85984	.408	215.1
	2	7	89821	.378	613.3
	3	10	85191	.316	552.5
	4	17	84624	.243	363.1
11	1	6	87911	.408	373.2
	2	5	88041	.447	501.1
	3	6	87911	.408	695.5
	4	5	87635	.447	323.9
	5	5	87693	.447	507.9

3. Statistical Analyses

Two OMNITAB programs were written to perform the statistical analyses, the ultimate purpose of which was to obtain overall weighted average values of the second and third law heats of sublimation and estimates of the uncertainties. The first OMNITAB program performed least-squares fits for each run to obtain the second law beats and the average third law heats. The program also made a preliminary test to detect laboratories that exhibited excessive scatter of points about the fitted curves (see sec. 6.1). The authors then examined the results and made tentative decisions regarding the data to be excluded from the weighted averages

and the estimated uncertainties [7]. The second OMNITAB program was then run to determine (1) the weighted average values of the second and third law heats of sublimation, (2) the uncertainties associated with the beats, and (3) the uncertainties expected for a typical in-control laboratory's measurements (see sec. 6.2). In the second OMNITAB program the rejected data were not used for the calculation of the weighted averages and standard deviations, but were used in all other statistical tests. This procedure avoids distorting the overall results, but still allows for further evaluation of all of the data.

The statistical analyses indicate the weighted

4

averages [8] and the associated standard deviations (standard errors) are as follows:

$A = -0.26 \pm 0.25$ cal $\cdot$ mol^{-1} $\cdot$ deg^{-1}

$B = $ 2nd law $\Delta H_{\text{sub298}} = 88,140 \pm 500$ cal $\cdot$ mol^{-1}

3rd law $\Delta H_{\text{sub298}} = 87,720 \pm 210$ cal $\cdot$ mol^{-1}.

The A coefficient is essentially zero which indicates the observed pressures and the free energy functions are in reasonable agreement. In the analyses it is tacitly assumed that the errors in the free energy functions are negligible. The second and third law ΔH_{sub298} are observed to be in good agreement. We believe the third law ΔH_{sub298} value of $87,720 \pm 210$ cal $\cdot$ mol^{-1} ($367,040 \pm 900$ J $\cdot$ mol^{-1}) [3] is to be preferred since its standard error is smaller and since the free energy functions for gold are believed reliable.

A laboratory measuring a single temperature-pressure curve may wish to compare its values with the weighted averages from this study. The total expected variance required for this comparison will be the sum of:

(1) the between-curve variance,

(2) the between-laboratory variance, and

(3) the variance of the weighted average.

Assembling the numerical values corresponding to these components of variance in the above order we obtain for the single curve case:

$V(A) \qquad = 0.020 f_1^2 \qquad + 0.0 \qquad + 0.063$

$V(\text{2nd law}) = 0.020 f_2^2 \qquad + 0.59 \times 10^6 \ + 0.24 \times 10^6$

$V(\text{3rd law}) = 0.070 \times 10^6 + 0.340 \times 10^6 + 0.046 \times 10^6$

where the f_1 and f_2 values may be calculated as indicated in section 6.1. The constants in these equations are based on pooled estimates of the variance components. By use of the variables, f_1 and f_2, allowance is made for the actual number of data points used and for the spread of $1/T$ values. The units of variance of the values are the squares of the units of the values being evaluated (A, B, or 3rd law heat). The numerical values in the variance equations have been derived using energy units of calories. The above formula giving the variance of the third law heat is based on the assumption that a laboratory makes at least five temperature-pressure measurements. For such a case, our analysis indicates the between-curve component of variance is approximately a constant, 0.070.

To illustrate the above equations for a "typical single curve case," assume that a laboratory measures a single temperature-pressure curve taking 11 points, one every 25 K, over the temperature range 1600 to 1850 K. For this case the f_1^2 and f_2^2 values

are calculated to be 43.2 and 1.28×10^8, respectively, and

$V(A) \qquad = 0.864 \qquad + 0 \qquad + 0.063$
$\qquad\qquad = 0.927$

$V(\text{2nd law}) = 2.56 \times 10^6 \ + 0.59 \times 10^6 \ + 0.24 \times 10^6$
$\qquad\qquad\quad = 3.39 \times 10^6 \qquad (3)$

$V(\text{3rd law}) = 0.070 \times 10^6 + 0.340 \times 10^6 + 0.046 \times 10^6$
$\qquad\qquad\quad = 0.456 \times 10^6 \qquad (4)$

The following limits, which are equal to twice the square root of the above variances, can be used for the estimation of maximum allowable differences between the single curve results obtained by the typical laboratory and the weighted averages. Approximately 95 percent of the time, a result obtained by the above described typical laboratory should fall within the following limits:

$A = 0.26 \pm 1.93$ cal $\cdot$ mol^{-1} $\cdot$ deg^{-1}

2nd law $= 87,720 \pm 3700$ cal $\cdot$ mol^{-1}

3rd law $= 87,720 \pm 1350$ cal $\cdot$ mol^{-1}.

Since the third law value is believed to be more accurate, we have replaced the second law weighted average by the third law weighted average, 87,720 cal $\cdot$ mol^{-1}.

A laboratory wishing to evaluate its own results should calculate its own specific f_1 and f_2 values for use in the above equations.

An examination of the values of the individual components of variance for the typical single curve case yields considerable information. For the second law case (eq 3) one can note that the between-curve variance is relatively large compared to the total variance (2.56/3.39). If a laboratory measures $(n-1)$ additional temperature-pressure curves the between-curve variance will be reduced to $(2.56 \times 10^6)/n$. For the third law case (eq 4) it can be observed that additional curves will not be particularly helpful since a large fraction of the total variance for the third law case is due to the between-laboratory component of variance (0.340/0.456).

By back calculating through the third law equation it is possible to determine approximate 95 percent limits for which the pressure-temperature relationship is known. This has been done using both the uncertainty of the weighted average third law value and the uncertainty of the typical single curve third law value. The results are shown in figure 6. From the weighted average third law limits it is seen that for a fixed temperature, the uncertainty in the associated pressure is approximately ± 13 percent, while for a fixed pressure, the uncertainty in the associated temperature is approximately ± 9 K. The corresponding limits for the typical single curve case are ± 45 percent and ± 30 K. The large uncertainty for the typical single curve case is primarily due to the large between-laboratory uncertainty.

5

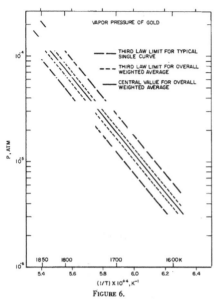

FIGURE 6.

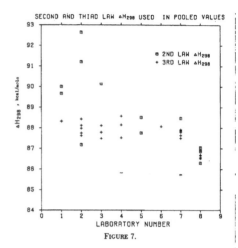

FIGURE 7.

A single laboratory's evaluation of error may be greatly underestimated if systematic between-laboratory errors are not considered. The diligent use of vapor pressure standard reference materials should help in the detection of elimination of such systematic errors.

4. Comparison of Second and Third Law Results

From the statistical analyses we have observed two fundamentally·different situations for the accepted second and third law results. Regarding the second law results, the between-curve and within laboratory variation was found to be no larger than that expected from the average scatter of temperature-pressure points about the curves. Furthermore, the second law case showed no (statistically) significant difference between the results from the different laboratories. For the third law case, however, a significant difference was found for both the between-curve and the between-laboratory results. It should be noted that the significant differences for the third law tests are due to the smaller third law uncertainties rather than to a wider spread of the values. The third law uncertainties are significantly smaller than the second law uncertainties. Figure 7 summarizes the accepted second and third law results.

List of Cooperating Laboratories

Aerospace Corporation, P. C. Marx, E. T. Chang, and N. A. Gokcen

Air Force Materials Laboratory (MAMS), H. L. Gegel
Air Force Materials Laboratory (MAYT), G. L. Haury
Douglas Advanced Research Laboratories, D. L. Hildenbrand
Gulf General Atomic, Inc., H. G. Staley, P. Winchell, J. H. Norman, and D. A. Bafus
Michigan State University, J. M. Haschke and H. A. Eick
National Bureau of Standards, E. R. Plante and A. B. Sessoms
Philco-Ford Corporation, N. D. Potter
Space Sciences, Inc., M. Farber, M. A. Frisch, and H. C. Ko
Universita Degli Studi di Roma, V. Piacenta and G. DeMaria
University of Pennsylvania, W. W. Worrell and A. Kulkarni

The authors are greatly indebted to the above listed cooperating laboratories for their vapor pressure measurements. D. L. Hildenbrand of Douglas Advanced Research Laboratories, should be given particular credit for the original impetus in the establishment of the vapor pressure standard reference materials program. The authors also wish to acknowledge aid received from the following NBS staff members. E. R. Plante has contributed freely to discussions dealing with thermodynamic aspects of the analysis, and J. J. Diamond has made an extensive vapor pressure literature survey from his information "Data Center on the Vaporization of Inorganic Materials." W. S. Horton contributed to discussions dealing with statistical procedures and F. L. McCrackin aided in the use of his computerized GRAPH routine.

5. Footnotes and References

[1] Standard reference material 745 may be ordered from the Office of Standard Reference Materials, National Bureau of Standards, Washington, D.C. 20234. This material is in the form of wire 1.4 mm (0.055 in) in diameter and 152 mm (6 in) long. The gold is homogeneous and 99.999 percent pure. The price for this material is $85 per unit; this includes a "Certificate of Analysis" containing specific recommendations for usage as well as several statistical tests by which a laboratory may evaluate its results.

[2] This $\Delta H_{sub\,298}$ value is in good agreement with the values 87,500 and 87,300 cal/mol quoted by:
Wagman, D. D., Evans, W. H., Parker, V. B., Halow, I., Bailey, S. M., and Schumn, R. H., Selected Values of Chemical Thermodynamic Properties, NBS Technical Note 270–4 (1969), and Hultgren, R., Orr, R. L., Anderson, P. D., and Kelley, K. K., Selected Values of Thermodynamic Properties of Metals and Alloys, pp 38–42, (June 1960), John Wiley & Sons, Inc., New York (1963).

[3] 1 calorie = 4.1840 joules
1 atmosphere = 101,325 newtons · meters^{-2}.

[4]

Temperature	Condensed phase [a]		Gas phase [b]	
	$\dfrac{G_T^\circ - H_{298}^\circ}{T}$		$\dfrac{G_T^\circ - H_{298}^\circ}{T}$	
K, (IPTS–68)	cal·mol^{-1}·deg^{-1}	(J·mol^{-1}·deg^{-1})[3]	cal·mol^{-1}·deg^{-1}	(J·mol^{-1}·deg^{-1})[3]
298.15	11.319	(47.359)	43.120	(180.414)
1200	15.352	(64.233)	46.304	(193.736)
1300	15.751	(65.902)	46.607	(195.004)
1338 (M.P.)	15.896	(66.509)	46.718	(195.468)
1400	16.236	(67.931)	46.894	(196.205)
1500	16.749	(70.078)	47.165	(197.338)
1600	17.233	(72.103)	47.426	(198.430)
1700	17.674	(73.948)	47.673	(199.464)
1800	18.117	(75.802)	47.910	(200.455)
1900	18.515	(77.467)	48.138	(201.409)
2000	18.913	(79.132)	48.356	(202.322)
2100	19.275	(80.647)	48.567	(203.204)
2200	19.636	(82.157)	48.768	(204.045)

[a] Converted to IPTS–68 from data of Tester, J. W., Feber, R. C., and Herrick, C. C., J. Chem. Eng. Data **13**, 419–21 (July 1968).
[b] From data of Hultgren, R., Orr, R. L., Anderson, P. D., and Kelley, K. K., Selected Values of Thermodynamic Properties of metals and Alloys, pp 38–42 (June 1960), John Wiley & Sons, Inc., New York (1963).

[5] A) Horton, W. S., J. Res. Nat. Bur. Stand. (U.S.) **70A** (Phys. and Chem.), No. 6, 533–9 (Nov.–Dec. 1966).
B) Cubicciotti, D., J. Phys. Chem. **70**, 2410–3 (1966).

[6] Hilsenrath, J., Ziegler, G. G., Messina, C. G., Walsh, P. J., and Herbold, R. J., OMNITAB, a Computer Program for Statistical and Numerical Analysis, Nat. Bur. Stand. (U.S.) Handb. 101, 256 pages, (March 1966).

[7] Preliminary examination of the data and the associated uncertainties for laboratories 9, 10, and 11 indicated that these laboratories deviated significantly from the consensus. An examination of the reports from the laboratories also indicated possible experimental difficulties. The results from these laboratories were, therefore, not included in further poolings. Subsequent statistical examination of these laboratories' data and associated uncertainties further indicated that these data should not be included in the pooled results.

[8] The results from eight cooperating laboratories with 27 curves and over 250 temperature-pressure points were used to determine these weighted averages. The weighting procedure used, is given in section 6.2.

[9] A special pooling procedure for standard deviations was used throughout this study. An example of the pooling procedure is as follows:

$$\text{pooled } S_{flt} = \frac{\sum_i \alpha_i S_{flt_i}}{\sum_i \beta_i}$$

where the sum is over the i curves and

$$\alpha_i = 2\nu_i + \frac{1}{2 + 3\nu_i}$$

$$\beta_i = 2\nu_i - \frac{1}{2} + \frac{2}{3 + 5\nu_i}$$

and ν_i = number of degrees of freedom. For a normal distribution, this pooling procedure for standard deviations will give results comparable to those obtained by the usual procedure of pooling variances. This procedure, however, has the advantage of being less sensitive to distortion by outlier values. The authors wish to thank B. L. Joiner of the National Bureau of Standards for the derivation of this pooling formula.

[10] Davies, O. L., Statistical Methods in Research and Production, pp. 97–99, Oliver and Boyd Publishers, London, England (1947).
[11] Mandel, J., The Statistical Analysis of Experimental Data, pp. 132–5, Interscience Publishers, New York (1964).

7

6. Appendix

This appendix gives additional details of the statistical analyses which were necessary for the evaluation of the gold vapor pressure data. The two-part statistical analysis has been made in terms of two large OMNITAB programs. An outline of the two parts is as follows:

6.1.

1. The temperature-pressure data for each run were given least-squares treatments described below to obtain the second and third law values and the associated uncertainties. In all fits each data point was given unit weight.
   A. For the second law equation, the least-squares model was $Y = A + BX$, where $X = 1/T$. The standard deviations of the coefficients (S_A and S_B) can be expressed in terms of the standard deviation of the fit (S_{fit}):

$$S_A = f_1 \cdot S_{fit} \qquad (5)$$

$$S_B = f_2 \cdot S_{fit} \qquad (6)$$

where

$$f_1 = \left[\frac{1}{N} + \frac{(\bar{X})^2}{\sum_i (X_i - \bar{X})^2} \right]^{1/2} \qquad (7)$$

$$f_2 = \left[\frac{1}{\sum_i (X_i - \bar{X})^2} \right]^{1/2} \qquad (8)$$

and $N =$ the number of data points. The f_1 and f_2 values provide a convenient quantitative description of the number and spread of the $X (= 1/T)$ values and have been used throughout the analyses. Since the OMNITAB least-squares fit program automatically gives the standard deviations for both the coefficients and the fit, the f_1 and f_2 values were conveniently calculated using eqs (5) and (6).
   B. The third law equation was also treated by least-squares. Here one obtains a single coefficient, C (the average of the individual ΔH_{sub298} values), the standard deviation of the coefficient (S_C), and the standard deviation of the individual ΔH_{sub298} values (S_{fit}). For the third law case, it can be shown that:

$$S_C = f_3 \cdot S_{fit} \qquad (9)$$

where

$$f_3 = \frac{1}{\sqrt{\# \text{ of points}}} \qquad (10)$$

It can be noted that eq (9) has the same form as eqs (5) and (6). The same general computational treatment was therefore used for both the second and third law results.

C. The results for the second and third law least-square fits are given in tables 3 and 4.
2. The authors next examined all results in terms of criteria A through F, listed below.
   A. The chi-square test. Comparisons were made of S_{fit} values from all curves.
      (1) A pooled $\bar{S}_{fit}$ was first calculated from the individual S_{fit} values from all curves [9].
      (2) Each individual S_{fit} was compared to the pooled $\bar{S}_{fit}$ using the approximate test:

$$\bar{S}_{fit} \sqrt{\frac{\chi^2_{\bar{\nu}, 0.025}}{\nu}} \leqslant S_{fit} \leqslant \bar{S}_{fit} \sqrt{\frac{\chi^2_{\bar{\nu}, 0.975}}{\nu}}$$

where χ^2 is the 0.025 or the 0.975 percentile of the chi-square distribution with ν degrees of freedom. A laboratory showing several curves for which the values of S_{fit} fell outside these two limits was noted for further evaluation.
      B. The between-curve (within-laboratory) differences for both the second and third law results.
      C. The overall differences between the second and third law results.
      D. The overall differences for results from the different laboratories.
      E. The possible drift of results with respect to time.
      F. The laboratory's experimental procedures.
3. Based on the above considerations, the data from laboratories 9, 10, and 11 were not used in further calculations of averages and pooled standard deviations. The results of laboratories 9, 10, and 11 were, however, compared to those of the other laboratories in the second OMNITAB program. This subsequent analysis confirmed the rejection decision. The variation of second law results for laboratories 10 and 11 was observed to be especially large. The results of laboratory 9 were not included because of combined minor difficulties in points C, E, and F above. The S_{fit} values for laboratories 7 and 8 were not used in further poolings since these laboratories did not randomly vary their temperature during the measurement of the temperature-pressure curves. As expected, the S_{fit} values for these laboratories were abnormally small. All other values from laboratories 7 and 8 were, however, used in the further calculations.

6.2.

1. In the second OMNITAB program, a comparison was made of runs within each laboratory. This comparison was made in terms of both the intercept A and the slope B for the curve fitted to each run, and in terms of the average third law heat derived from each run. Using the F test, the variance of the A values between curves within each laboratory was compared to the estimate of this variance derived from the pooled

8

S_{fit}. The B and the third law heat values were similarly treated.

2. A comparison was made of laboratories with each other. First, a pooled value was obtained for the between curves (within laboratories) standard deviation for each of the three parameters A, B, and third law heat; an average value (for each of the three parameters) was also computed. Then, using Student's t test, the deviation of the average value of each laboratory from the overall weighted average was compared to the pooled standard deviation between curves (within laboratories). In this way, detailed information was obtained on the variability between laboratories in terms of the deviation of each individual laboratory from the consensus value.

3. An analysis of variance was made [10] for each of the three parameters, A, B, and third law heat, using the estimated values of these parameters accepted after application of the first OMNITAB program. The purpose of the analysis was to estimate the components of the within- and between-laboratory variance.

4. Overall weighted average (A, B, and third law heat) values and the associated variances were determined. Since the laboratories did not submit the same number of runs, the overall weighted averages are dependent on the specific weighting procedure used. Statistically, a proper weighting procedure would be one that minimizes the variance of the weighted average. The weighting factors obtained by this procedure are functions of the ratio of the between- to within-laboratory components of variance. Denoting the ratio for A by ρ, it can be shown that laboratory i with n_i curves has the weighting factor:

$$W_i = \frac{n_i}{1 + n_i \rho}.$$

The value of ρ can be estimated from the results of the analysis of variance [10]. The weighted average $\bar{A}$ will be:

$$\bar{A} = \frac{\sum_i W_i \bar{A}_i}{\sum_i W_i}$$

where $\bar{A}_i$ is the average A value for laboratory i.

Using this procedure, the variance of $\bar{A}$ will be smaller than for any other weighting procedure, and its approximate value will be:

$$V(\bar{A}) = \frac{\text{``}A\text{'' component of within-lab. variance [11]}}{\sum_i W_i}.$$

The values for the B and the third law heat were evaluated in an analogous manner using the ρ and n_i values corresponding to these parameters.

Two extreme cases for the weighting factor deserve special attention. For the situation where the ratio, ρ, of the between- to within-laboratory components of variance is large with respect to unity, essentially equal weight is given to each laboratory. For the situation where the ratio ρ is close to zero, each curve is given essentially equal weight. The ρ values for A, B, and third law heat which we obtained from the analysis of variance are 0.0, 0.169, and 4.835, respectively.

5. Finally, the components of variance were assembled from the analysis of variance to estimate the uncertainties for the pooled and single curve values.

9

6.3. Experimental Data

TABLE 2. *List of Experimental Temperature-Pressure Data*

Lab 1, Run 1		Lab 1, Run 2		Lab 2, Run 1		Lab 2, Run 2		Lab 2, Run 3	
T, K	P, ATM	T, K	P, ATM	T, K	P, ATM	T, K	P, ATM	T, K	P, ATM
1796.2	6.520×10^{-5}	1753.1	3.330×10^{-5}	1680.0	1.110×10^{-5}	1689.0	1.490×10^{-5}	1855.3	1.440×10^{-4}
1894.4	1.960×10^{-4}	1653.9	8.170×10^{-6}	1724.0	1.960×10^{-5}	1739.1	3.470×10^{-5}	1912.4	2.740×10^{-4}
1727.0	2.540×10^{-5}	1906.4	2.410×10^{-4}	1780.2	4.890×10^{-5}	1804.2	7.790×10^{-5}		
1840.3	1.130×10^{-4}	1694.0	1.680×10^{-5}	1821.2	7.310×10^{-5}	1865.3	1.840×10^{-4}		
1964.5	4.000×10^{-4}	1556.7	1.380×10^{-6}	1871.3	1.410×10^{-4}	1900.4	2.470×10^{-4}		
1705.0	1.760×10^{-5}	1847.3	1.040×10^{-4}	1900.4	1.960×10^{-4}	1927.5	3.530×10^{-4}		
1998.6	6.040×10^{-4}	1613.8	3.920×10^{-6}	1860.3	1.240×10^{-4}	1845.3	1.420×10^{-4}		
1924.4	2.790×10^{-4}	1972.5	5.510×10^{-4}	1793.2	5.610×10^{-5}	1774.1	3.870×10^{-5}*		
1635.9	5.880×10^{-6}	1584.8	2.920×10^{-6}	1747.1	3.260×10^{-5}	1673.9	1.270×10^{-5}		
1764.1	3.560×10^{-5}	1935.5	3.010×10^{-4}	1703.0	1.780×10^{-5}	1641.9	7.420×10^{-6}		
1677.0	9.150×10^{-6}	1820.2	7.560×10^{-5}						
		1772.1	3.700×10^{-5}						

*This point discarded.

Lab 2, Run 4		Lab 2, Run 5		Lab 3, Run 1		Lab 3, Run 2		Lab 3, Run 3	
T, K	P, ATM	T, K	P, ATM	T, K	P, ATM	T, K	P, ATM	T, K	P, ATM
1690.0	1.440×10^{-5}	1712.0	2.060×10^{-5}	1746.1	3.310×10^{-5}	1686.0	1.350×10^{-5}	1578.8	2.900×10^{-6}
1727.0	2.800×10^{-5}	1776.1	4.650×10^{-5}	1684.0	1.480×10^{-5}	1729.1	2.450×10^{-5}	1629.9	7.000×10^{-6}
1675.9	1.260×10^{-5}			1748.1	3.520×10^{-5}	1631.9	6.600×10^{-6}	1681.0	1.540×10^{-5}
1651.9	9.180×10^{-6}			1794.2	6.680×10^{-5}	1686.0	1.360×10^{-5}	1591.8	3.800×10^{-6}
1614.8	4.830×10^{-6}			1785.2	5.620×10^{-5}	1720.0	2.230×10^{-5}	1607.8	5.100×10^{-6}
1599.8	3.450×10^{-6}			1798.2	7.010×10^{-5}	1756.1	3.500×10^{-5}	1659.9	1.150×10^{-5}
				1753.1	4.030×10^{-5}	1803.2	6.500×10^{-5}	1649.9	1.030×10^{-5}
				1707.0	1.990×10^{-5}	1735.1	2.780×10^{-5}	1604.8	5.200×10^{-6}
				1655.9	9.900×10^{-6}	1659.9	1.020×10^{-5}		
				1673.9	1.370×10^{-5}	1591.8	3.300×10^{-6}		
				1621.8	6.000×10^{-6}				

TABLE 2. *List of Experimental Temperature-Pressure Data*—Continued

Lab 4, Run 1		Lab 4, Run 2		Lab 4, Run 3		Lab 5, Run 1		Lab 5, Run 2	
T, K	P, ATM	T, K	P, ATM	T, K	P, ATM	T, K	P, ATM	T, K	P, ATM
2022.6	8.209×10^{-4}	1719.0	2.049×10^{-5}	1719.0	2.316×10^{-5}	1509.6	9.960×10^{-7}	1736.1	3.749×10^{-5}
1698.0	1.592×10^{-5}	1883.4	1.568×10^{-4}	1722.0	2.895×10^{-5}	1563.7	2.600×10^{-6}	1784.2	6.236×10^{-5}
1758.1	3.237×10^{-5}	1824.2	9.237×10^{-5}	1866.3	1.566×10^{-4}	1619.8	6.590×10^{-6}	1814.2	9.504×10^{-5}
1857.3	1.329×10^{-4}	1912.4	2.149×10^{-4}	1926.4	3.934×10^{-4}	1667.9	1.410×10^{-5}	1772.1	5.780×10^{-5}
1951.5	4.793×10^{-4}	1778.2	4.074×10^{-5}	1820.2	9.684×10^{-5}	1812.2	1.000×10^{-4}	1759.1	4.841×10^{-5}
1879.4	1.637×10^{-4}	1957.5	3.625×10^{-4}	1768.1	4.816×10^{-5}			1719.0	2.903×10^{-5}
1785.2	6.039×10^{-4}	2025.6	6.801×10^{-4}	1915.4	3.408×10^{-4}			1662.9	1.230×10^{-5}
1774.1	4.645×10^{-5}			2009.6	7.816×10^{-4}			1635.9	7.725×10^{-6}
2010.6	6.486×10^{-4}			1973.5	5.934×10^{-4}			1583.8	3.740×10^{-6}
1904.4	2.058×10^{-4}							1536.7	1.811×10^{-6}
								1502.6	9.439×10^{-7}
								1453.5	3.597×10^{-7}
								1471.5	5.139×10^{-7}
								1523.6	1.270×10^{-6}
								1581.8	3.732×10^{-6}
								1630.9	7.429×10^{-6}
								1663.9	1.439×10^{-5}
								1707.0	2.618×10^{-5}
								1738.1	4.236×10^{-5}
								1773.1	6.250×10^{-5}
								1832.3	1.324×10^{-4}
								1815.2	1.042×10^{-4}
								1811.2	9.395×10^{-5}
								1786.2	7.584×10^{-5}
								1764.1	5.406×10^{-5}
								1712.0	2.824×10^{-5}
								1681.0	1.636×10^{-5}
								1633.9	7.707×10^{-6}
								1598.8	4.334×10^{-6}
								1548.7	1.949×10^{-6}
								1488.6	6.791×10^{-7}

Lab 6, Run 1		Lab 7, Run 1		Lab 7, Run 2		Lab 7, Run 3		Lab 7, Run 4	
T, K	P, ATM	T, K	P, ATM	T, K	P, ATM	T, K	P, ATM	T, K	P, ATM
1323.2	1.780×10^{-8}	1499.6	7.024×10^{-7}	1506.6	8.129×10^{-7}	1506.6	8.160×10^{-7}	1534.7	1.386×10^{-6}
1323.7	1.600×10^{-8}	1506.6	7.365×10^{-7}	1535.7	1.354×10^{-6}	1536.7	1.400×10^{-6}	1564.7	2.309×10^{-6}
1325.2	1.320×10^{-8}	1541.7	1.442×10^{-6}	1585.8	3.063×10^{-6}	1568.7	2.295×10^{-6}	1584.8	3.210×10^{-6}
1325.7	1.790×10^{-8}	1544.7	1.534×10^{-6}	1605.8	4.279×10^{-6}	1597.8	3.704×10^{-6}	1603.8	4.383×10^{-6}
		1564.7	2.199×10^{-6}	1627.9	5.852×10^{-6}	1615.8	4.977×10^{-6}	1621.8	5.867×10^{-6}
		1577.8	2.799×10^{-6}	1636.9	6.797×10^{-6}	1632.9	6.626×10^{-6}	1630.9	6.795×10^{-6}
		1599.8	3.844×10^{-6}					1501.6	6.961×10^{-7}
		1602.8	4.117×10^{-6}					1540.7	1.460×10^{-6}
		1617.8	5.306×10^{-6}					1569.7	2.418×10^{-6}
		1632.9	6.430×10^{-6}					1589.8	3.336×10^{-6}
								1607.8	4.548×10^{-6}
								1626.8	6.205×10^{-6}

11

TABLE 2. *List of Experimental Temperature-Pressure Data* — Continued

Lab 7, Run 5		Lab 7, Run 6		Lab 7, Run 7		Lab 8, Run 1		Lab 8, Run 2	
T, K	P, ATM	T, K	P, ATM	T, K	P, ATM	T, K	P, ATM	T, K	P, ATM
1448.5	2.717×10^{-7}	1591.8	3.564×10^{-6}	1569.7	2.578×10^{-6}	1444.5	3.480×10^{-7}	1555.7	2.980×10^{-6}
1488.6	6.068×10^{-7}	1603.8	4.195×10^{-6}	1643.9	8.478×10^{-6}	1478.6	7.170×10^{-7}	1582.8	4.730×10^{-6}
1527.7	1.261×10^{-6}	1506.6	8.250×10^{-7}	1684.0	1.659×10^{-5}	1510.1	1.280×10^{-6}	1610.8	7.270×10^{-6}
1563.7	2.421×10^{-6}	1545.7	1.608×10^{-6}	1710.0	2.404×10^{-5}	1544.7	2.360×10^{-6}	1642.9	1.176×10^{-5}
1582.8	3.238×10^{-6}	1564.7	2.239×10^{-6}	1606.8	4.892×10^{-6}	1575.7	3.950×10^{-6}	1665.9	1.635×10^{-5}
1600.8	4.249×10^{-6}	1584.8	3.076×10^{-6}	1655.9	1.129×10^{-5}	1605.8	6.450×10^{-6}	1692.0	2.342×10^{-5}
		1604.8	4.206×10^{-6}	1690.0	1.758×10^{-5}	1639.9	1.087×10^{-5}	1719.0	3.380×10^{-5}
		1614.8	4.920×10^{-6}			1673.9	1.785×10^{-5}	1746.1	5.000×10^{-5}
		1604.8	4.275×10^{-6}			1702.0	2.663×10^{-5}	1772.1	7.154×10^{-5}
		1611.8	4.898×10^{-6}			1725.5	3.692×10^{-5}	1798.2	1.009×10^{-4}
		1546.7	1.628×10^{-6}			1737.1	4.335×10^{-5}	1825.2	1.440×10^{-4}
		1505.6	7.955×10^{-7}						
		1545.7	1.595×10^{-6}						
		1564.7	2.222×10^{-6}						
		1584.8	3.024×10^{-6}						

Lab 8, Run 3		Lab 8, Run 4		Lab 8, Run 5		Lab 9, Run 1		Lab 9, Run 2	
T, K	P, ATM	T, K	P, ATM	T, K	P, ATM	T, K	P, ATM	T, K	P, ATM
1448.4	3.900×10^{-7}	1439.5	3.140×10^{-7}	1562.7	3.270×10^{-6}	1615.8	1.140×10^{-5}	1738.1	7.260×10^{-5}
1480.6	7.600×10^{-7}	1483.6	7.780×10^{-7}	1589.8	5.020×10^{-6}	1528.7	2.850×10^{-6}	1631.9	1.520×10^{-5}
1508.5	1.272×10^{-6}	1515.6	1.370×10^{-6}	1618.8	8.150×10^{-6}	1397.4	2.180×10^{-7}	1518.6	2.140×10^{-6}
1542.2	2.320×10^{-6}	1548.2	2.440×10^{-6}	1644.9	1.210×10^{-5}	1504.6	1.830×10^{-6}	1411.4	2.820×10^{-7}
1564.3	3.365×10^{-6}	1579.8	4.230×10^{-6}	1673.9	1.850×10^{-5}	1631.9	1.660×10^{-5}	1547.7	3.440×10^{-6}
1592.7	5.350×10^{-6}	1615.3	7.370×10^{-6}	1701.0	2.718×10^{-5}	1437.5	5.200×10^{-7}	1651.9	1.900×10^{-5}
1621.8	8.446×10^{-6}	1654.4	1.398×10^{-5}	1727.0	3.894×10^{-5}	1764.1	1.050×10^{-4}	1708.0	4.550×10^{-5}
1654.9	1.386×10^{-5}	1684.0	2.013×10^{-5}	1754.1	5.587×10^{-5}	1604.8	1.060×10^{-5}	1597.8	7.260×10^{-6}
1654.5	1.366×10^{-5}	1709.5	2.900×10^{-5}	1780.2	7.915×10^{-5}	1710.0	4.990×10^{-5}	1468.5	9.840×10^{-7}
		1735.1	4.110×10^{-5}	1800.2	1.030×10^{-4}	1572.7	5.930×10^{-6}	1567.7	4.580×10^{-6}
				1827.2	1.448×10^{-4}	1739.1	7.430×10^{-5}	1442.5	5.450×10^{-7}
						1483.6	1.310×10^{-6}	1666.9	2.640×10^{-5}
						1687.0	3.470×10^{-5}	1754.1	9.490×10^{-5}
								1785.2	1.250×10^{-4}

Lab 9, Run 3		Lab 9, Run 4		Au₂ Lab 9, Run 1		Au₂ Lab 9, Run 2		Au₂ Lab 9, Run 3	
T, K	P, ATM	T, K	P, ATM	T, K	P, ATM	T, K	P, ATM	T, K	P, ATM
1754.1	9.400×10^{-5}	1744.1	6.610×10^{-5}	1682.0	6.320×10^{-8}	1779.2	3.420×10^{-7}	1805.2	5.220×10^{-7}
1657.9	1.910×10^{-5}	1646.9	1.690×10^{-5}	1764.1	3.130×10^{-7}	1692.0	7.070×10^{-8}	1718.0	1.150×10^{-7}
1532.7	2.410×10^{-6}	1528.7	2.150×10^{-6}	1708.0	1.110×10^{-7}	1646.9	2.190×10^{-8}	1661.9	2.840×10^{-8}
1442.5	3.930×10^{-7}	1437.5	3.500×10^{-7}	1820.2	7.830×10^{-7}	1754.1	2.130×10^{-7}	1738.1	1.570×10^{-7}
1561.7	4.060×10^{-6}	1566.7	3.470×10^{-6}	1744.1	2.090×10^{-7}	1677.0	3.550×10^{-8}	1683.0	4.410×10^{-8}
1677.0	2.710×10^{-5}	1677.0	2.500×10^{-5}	1795.2	5.120×10^{-7}	1728.1	9.270×10^{-8}	1780.2	2.980×10^{-7}
1738.1	6.860×10^{-5}	1733.1	4.930×10^{-5}	1656.9	4.020×10^{-8}	1805.2	5.820×10^{-7}	1826.2	7.360×10^{-7}
1622.8	1.250×10^{-5}	1621.8	9.080×10^{-6}						
1508.6	1.440×10^{-6}	1502.6	1.040×10^{-6}						
1406.4	1.980×10^{-7}	1400.4	1.570×10^{-7}						
1597.8	7.530×10^{-6}	1592.8	6.590×10^{-6}						
1478.6	8.460×10^{-7}	1473.5	7.590×10^{-7}						
1718.0	4.740×10^{-5}	1708.0	3.880×10^{-5}						
1800.2	1.380×10^{-4}	1785.2	1.280×10^{-4}						

12

TABLE 2. *List of Experimental Temperature-Pressure Data*—Continued

Au₂ Lab 9, Run 4		Lab 10, Run 1		Lab 10, Run 2		Lab 10, Run 3		Lab 10, Run 4	
T, K	P, ATM	T, K	P, ATM	T, K	P, ATM	T, K	P, ATM	T, K	P, ATM
1785.2	4.050×10^{-7}	1692.0	2.800×10^{-5}	1641.9	$1.470 \times 10^{-5*}$	1717.0	3.700×10^{-5}	1656.9	2.400×10^{-5}
1713.0	9.110×10^{-8}	1651.9	1.470×10^{-5}	1792.2	4.800×10^{-5}	1712.0	3.700×10^{-5}	1707.0	4.800×10^{-5}
1661.9	2.560×10^{-8}	1631.9	1.210×10^{-5}	1792.2	4.200×10^{-5}	1651.9	2.100×10^{-5}	1646.9	2.200×10^{-5}
1739.1	1.240×10^{-7}	1571.7	5.000×10^{-6}	1757.1	1.980×10^{-5}	1611.8	1.250×10^{-5}	1606.8	1.290×10^{-5}
1688.0	5.140×10^{-8}	1546.7	3.100×10^{-6}	1697.0	1.000×10^{-5}	1576.7	6.500×10^{-6}	1571.7	7.100×10^{-6}
1774.1	2.710×10^{-7}	1621.8	9.400×10^{-6}	1656.9	5.400×10^{-6}	1536.7	3.700×10^{-6}	1541.7	4.300×10^{-6}
1815.2	5.900×10^{-7}			1646.9	4.200×10^{-6}	1511.6	2.300×10^{-6}	1501.6	2.200×10^{-6}
				1702.0	8.200×10^{-6}	1571.7	7.300×10^{-6}	1521.6	3.100×10^{-6}
						1671.9	2.700×10^{-5}	1606.8	1.400×10^{-5}
						1511.6	2.100×10^{-6}	1651.9	2.900×10^{-5}
								1727.0	7.100×10^{-5}
								1656.9	2.900×10^{-5}
								1621.8	1.270×10^{-5}
								1571.7	6.100×10^{-6}
								1536.7	3.700×10^{-6}
								1486.6	1.910×10^{-6}
								1541.7	3.500×10^{-6}
								1586.8	$5.800 \times 10^{-6*}$

*This point discarded.

Lab 11, Run 1		Lab 11, Run 2		Lab 11, Run 3		Lab 11, Run 4		Lab 11, Run 5	
T, K	P, ATM	T, K	P, ATM	T, K	P, ATM	T, K	P, ATM	T, K	P, ATM
1536.7	1.150×10^{-6}	1514.6	8.130×10^{-7}	1555.7	2.660×10^{-6}	1578.8	3.180×10^{-6}	1513.6	7.880×10^{-7}
1571.7	2.160×10^{-6}	1539.7	1.040×10^{-6}	1594.8	3.420×10^{-6}	1585.8	3.350×10^{-6}	1586.8	4.030×10^{-6}
1584.8	2.950×10^{-6}	1554.7	1.950×10^{-6}	1634.9	7.160×10^{-6}	1616.8	6.220×10^{-6}	1596.8	4.610×10^{-6}
1634.9	7.190×10^{-6}	1567.7	2.420×10^{-6}	1637.9	7.740×10^{-6}	1625.8	6.410×10^{-6}	1632.9	6.820×10^{-6}
1662.9	1.170×10^{-5}	1622.8	6.030×10^{-6}	1690.0	1.350×10^{-5}	1639.9	6.830×10^{-6}	1682.0	1.320×10^{-5}
1677.0	1.440×10^{-5}			1714.0	1.670×10^{-5}				

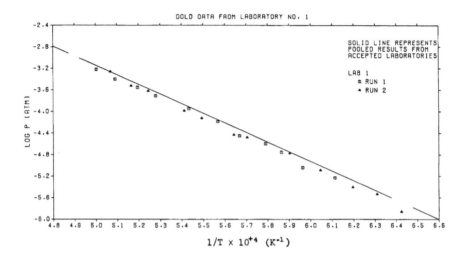

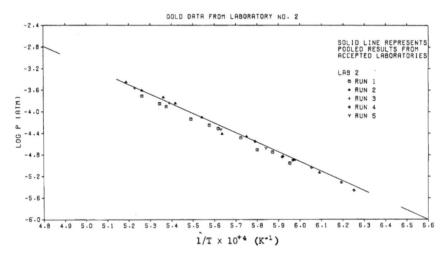

FIGURE 1.

14

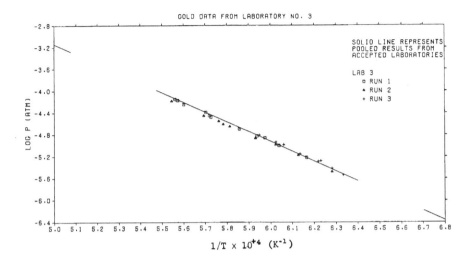

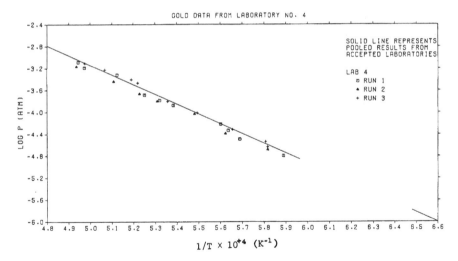

FIGURE 2.

15

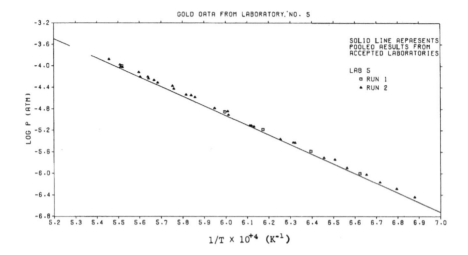

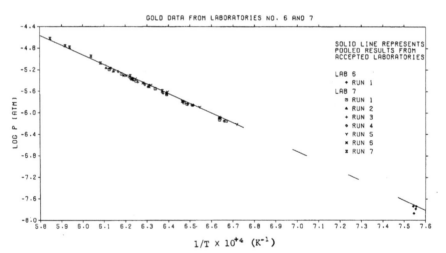

FIGURE 3.

16

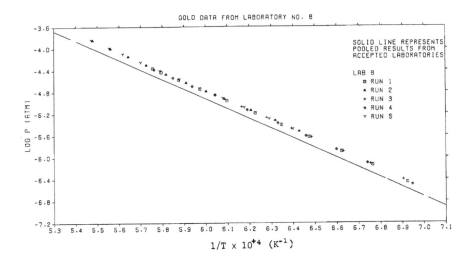

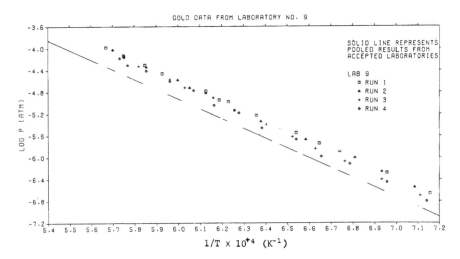

FIGURE 4.

17

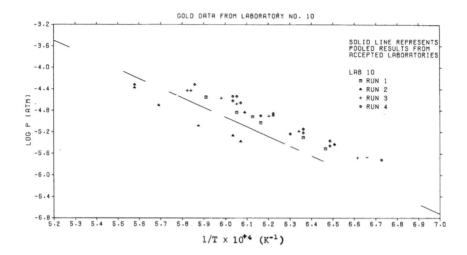

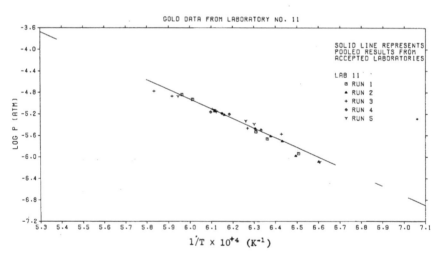

FIGURE 5.

18

NBS TECHNICAL PUBLICATIONS

PERIODICALS

JOURNAL OF RESEARCH reports National Bureau of Standards research and development in physics, mathematics, chemistry, and engineering. Comprehensive scientific papers give complete details of the work, including laboratory data, experimental procedures, and theoretical and mathematical analyses. Illustrated with photographs, drawings, and charts.

Published in three sections, available separately:

● **Physics and Chemistry**

Papers of interest primarily to scientists working in these fields. This section covers a broad range of physical and chemical research, with major emphasis on standards of physical measurement, fundamental constants, and properties of matter. Issued six times a year. Annual subscription: Domestic, $9.50; foreign, $11.75*.

● **Mathematical Sciences**

Studies and compilations designed mainly for the mathematician and theoretical physicist. Topics in mathematical statistics, theory of experiment design, numerical analysis, theoretical physics and chemistry, logical design and programming of computers and computer systems. Short numerical tables. Issued quarterly. Annual subscription: Domestic, $5.00; foreign, $6.25*.

● **Engineering and Instrumentation**

Reporting results of interest chiefly to the engineer and the applied scientist. This section includes many of the new developments in instrumentation resulting from the Bureau's work in physical measurement, data processing, and development of test methods. It will also cover some of the work in acoustics, applied mechanics, building research, and cryogenic engineering. Issued quarterly. Annual subscription: Domestic, $5.00; foreign, $6.25*.

TECHNICAL NEWS BULLETIN

The best single source of information concerning the Bureau's research, developmental, cooperative and publication activities, this monthly publication is designed for the industry-oriented individual whose daily work involves intimate contact with science and technology—*for engineers, chemists, physicists, research managers, product-development managers, and company executives.* Annual subscription: Domestic, $3.00; foreign, $4.00*.

* Difference in price is due to extra cost of foreign mailing.

NONPERIODICALS

Applied Mathematics Series. Mathematical tables, manuals, and studies.

Building Science Series. Research results, test methods, and performance criteria of building materials, components, systems, and structures.

Handbooks. Recommended codes of engineering and industrial practice (including safety codes) developed in cooperation with interested industries, professional organizations, and regulatory bodies.

Special Publications. Proceedings of NBS conferences, bibliographies, annual reports, wall charts, pamphlets, etc.

Monographs. Major contributions to the technical literature on various subjects related to the Bureau's scientific and technical activities.

National Standard Reference Data Series. NSRDS provides quantitive data on the physical and chemical properties of materials, compiled from the world's literature and critically evaluated.

Product Standards. Provide requirements for sizes, types, quality and methods for testing various industrial products. These standards are developed cooperatively with interested Government and industry groups and provide the basis for common understanding of product characteristics for both buyers and sellers. Their use is voluntary.

Technical Notes. This series consists of communications and reports (covering both other agency and NBS-sponsored work) of limited or transitory interest.

Federal Information Processing Standards Publications. This series is the official publication within the Federal Government for information on standards adopted and promulgated under the Public Law 89–306, and Bureau of the Budget Circular A–86 entitled, Standardization of Data Elements and Codes in Data Systems.

CLEARINGHOUSE

The Clearinghouse for Federal Scientific and Technical Information, operated by NBS, supplies unclassified information related to Government-generated science and technology in defense, space, atomic energy, and other national programs. For further information on Clearinghouse services, write:

Clearinghouse
U.S. Department of Commerce
Springfield, Virginia 22151

Order NBS publications from: Superintendent of Documents
Government Printing Office
Washington, D.C. 20402

U.S. DEPARTMENT OF COMMERCE
WASHINGTON, D.C.   20230

———

OFFICIAL BUSINESS

POSTAGE AND FEES PAID
U.S. DEPARTMENT OF COMMERCE

CPSIA information can be obtained
at www.ICGtesting.com
Printed in the USA
LVHW021507261118
598291LV00012B/1240